# PRIME

# PATTERNS

## *Neoteric Patterns in the Prime Numbers*

# JOYCE P. BOWEN, PhD

*Building Blocks of the Prime Numbers Have Been Identified*

Cover Design by Yaoundé Olu

The author may be contacted at:

Email: uniphysics@aol.com

www.astropoint.net

Astropoint Research

Chicago, Illinois

# TABLE OF CONTENTS

# INTRODUCTION

Prime numbers are numbers that have only the factor of themselves and one. They are enigmatic, in that no discernable pattern of organization is evident. When Einstein said God doesn't play dice when expressing exasperation about quantum physics, he could have said the same thing about prime numbers.

There have been some advances in understanding them. Researchers have been able to make progress regarding the conjecture that there are an infinite number of pairs of prime numbers that differ only by two. It has also been found that the frequency of twin primes decreases as one gets to larger numbers. Brun's constant, another revealing aspect of their nature, is found by adding the reciprocals of successive twin primes. When this is done, the sum converges to a specific numerical value which is 1.902160582310… The reciprocals of all of the primes, on the other hand, diverge.

The Goldbach Conjecture, that every even number is the sum of two primes, has not yet been proven. It is known, however, that all twin primes are of the multiple 6k +/- 1.

Other mathematicians who have thrown their hats into the prime ring are Mersenne (Mersenne Primes), Sophie Germaine (Sophie Germaine Primes) and Fermat (Fermat Primes) just to name a few. One of the most intriguing of prime conjectures is the Riemann Hypothesis. Briefly stated, when studying the distribution of prime numbers, Riemann extended Euler's zeta function, defined for real part greater than one, to the entire complex plane. When he did, he noted that his zeta function had trivial zeros at $-2$, $-4$, $-6$,… and that all nontrivial zeros were symmetric about the line $Re(s) + \frac{1}{2}$. The hypothesis is that all nontrivial zeros are on this line. The Clay

Mathematics Institute has offered to pay $1 million dollars to anyone that can prove this hypothesis to be true.

It has been discovered, however, that there is another way to look at prime numbers. *Prime Patterns* utilizes Increment Analysis, a Unimath technique of intra-space analysis, to unravel hidden patterns in numerical sequences. This book provides the details for a new way to assess prime numbers and has revealed a pattern that has been under our noses the whole time.

The second half of this book will utilize Unimath and its main tool, Increment Analysis, to reveal a new way to assess numbers and apply it to discover even more patterns in the prime numbers as well as in numbers in general.

# UNIMATH

In today's society numbers are merely seen as tools for calculation. The notion that numbers are things in themselves and have deeper import is not always considered. In this book I will examine numbers with the goal of uncovering deeper meanings and evidence of an underlying order to seemingly random numerical systems. Unimath, which uses Increment Analysis to assess number relationships between adjacent numbers in numerical sequences, is a technique used to identify and synthesize number relationships. Most of this information appeared in my previous book, ***The Law of Digit Balance***.

*Increment Analysis differs from the calculus of sequence differences in that the result of analysis is <u>a charged number line</u>.  Increment Analysis reveals previously unobserved patterns in Fibonacci numbers, Lucas numbers and prime numbers. In a nutshell, Increment Analysis reveals obvious, arithmetical characteristics of number sequences that were heretofore hidden, and where the increments of all whole numbers, or fragments of irrational or transcendental numbers sum to zero while exhibiting beautiful symmetries.*[1]

There are mirror image groupings that cancel each other with a resultant of zero in circular numbers. ***In some cases, 58 and 60 element mirror image sequences are revealed.*** In addition, there are oscillating digits denoting a wave pattern whose sum tends to converge to zero in non-circular number sequences. When zeros are added before and after ***any sequence***, the result is digits which cancel to zero. Increment Unimath, therefore, uncovers a type of ***numerical homeostasis with zero as a focal point.*** In other words, there seems to be a ***Law of Digit Balance*** *wherein the digits of all numbers in all number sequences have increments that sum to zero in beautifully arranged symmetries.*

---

1 See the appendix for further information about the "gender" of number sequences.

# RULES OF ANALYSIS

1. Select a number sequence and evaluate the increments between adjacent numbers so that they become positive or negative in relation to each other.  Ignore decimal points and treat the number as a continuous sequence. For example, the reciprocal of seven is a series of repeating numbers 0.1428571…:

2.

$$1 \quad 4 \quad 2 \quad 8 \quad 5 \quad 7 \quad 1...:$$

**The increment between 1 and 4 is  +3**

**"          "          "     4 and 2 is   −2**

**"          "          "     2 and 8 is   +6**

**"          "          "     8 and 5 is   −3**

**"          "          "     5 and 7 is   +2**

**"          "          "     7 and 1 is    -6**

**The resulting number sequence, therefore, is:**

**+3  -2  +6  -3  +2  -6…**

**The resulting Binaric is:**

**+11   -11 = 0**

3. Isolate the resulting binary patterns and cancel them to zero, leaving one or more remainders. In the above case,  +3  -2  and −3 +2 cancel each other,  and the +6 and −6 also cancel each other, resulting in 0.
4. To see the underlying patterns of any sequence or sequence segment that is not a circular number, you add zeros before and after the sequence and proceed with the methodology outlined in number 1 above. The digits will always cancel to zero.
5. Though it is important to keep in mind that each digit is an increment that is ten times more precise than the last in the decimal expansion of numbers, the numbers are taken at face value.

# POSTULATE

*The increments between adjacent numbers in circular,[2] or revolving number sequences generate oscillating mirror image number groupings that always cancel to 0.*

*EXPERIMENTAL DATA*

**Let n = digits in a number sequence**

**Let $\Delta$ = the increments between adjacent numbers in the sequence**

**If n is extended indefinitely,**

## Then:

$$\Sigma \, \Delta \, n = 0$$

---

2 Circular numbers produce the same arithmetic sequence of numbers, but starting in different positions, when multiplied or divided by given numbers.

# PHANTOM VALUES

*The resulting numerical binarics from the assessment of a number sequence are called Phantom Values.*

Phantom Values (PhV) are the increments between numbers. For example, in the following example, the number sequence has the following phantom values.

**Number: 1/63 =**

**Sequence:  0.015873 0…**

**Phantom Values: +1  +4  +3    -1  - 4  -3**

**…which is +8   -8**

**The reciprocal of the number thirteen, 1/13, is**

**0.0769230769230…**

**The resulting sequence is:**

**+7 -1 +3   -7 +1 –3   +7 -1 +3**

**-7 +1 - 3…∞ = 0**

**The above resulting binaric: +11   -11**

**The reciprocal of the number seventeen =**

1/17=0.05882352941176470

+5 +3  0 - 6    +1 +2 - 3   +7

- 5 - 3  0 +6    -1 -2 +3   -7...∞ = 0

**The above resulting binaric: +27  -27**

The reciprocal of the number 47:

1/47 = 0.02127659574468085106382978723404 25

+2 -1 +1 +5 -1 -1 +4 -4 +2 –3 0 +2 +2
-8 +8 -3 -4 -1 +6
 -3 +5 -6 +7

-2 +1 -1 -5 +1 +1 -4 +4 -2 +3 0 -2 -2
+8 -8 +3 +4 +1 -6
+3 –5 +6 -7

**Resulting binaric: +79  -79**

# THE MOTHER BINARIC

The binary that underlies the numbers zero through nine is the Mother Binaric, +9 -9 shown below.

+1+1 +1  +1 +1 +1 +1   +1 +1  -9

0  1  2  3   4   5   6   7   8  9  0 =

**+9  -9**

# GENERATION OF CONTIGUOUS PRIME NUMBERS

# CONJECTURE

**Prime numbers can be generated by addressing the increments between prime numbers in CONTIGUOUS primes in the following manner:**

$$P_1 + \Delta \mid \Delta - 1 + \frac{1}{\Phi} = P_2$$

*...this reads Prime 1 plus increment or increment minus 1 plus the reciprocal of Phi equals Prime 2, the next contiguous prime*

**METHODOLOGY:** The building blocks of the prime numbers is Phi, or the reciprocal of Phi. The Increment between contiguous primes is added to the reciprocal of Phi, 0.6180 where the zero is replaced by the increment number *or one less than the Increment number in every case*. For example, if the increment is 3, the number **3**.6180 is added to the previous prime to generate the next prime. The following chart assesses increments and generates primes through 127.

# GENERATION OF THE PRIME NUMBERS THROUGH 127

| P1 (PRIME 1) | P2 (PRIME 2) | INCREMENT Add this number, *or one less*, to .6180 to generate the Increment ➡ | INCREMENT Generated from previous raw prime $\Phi^3$ | RAW PRIME | RAW PRIME ROUNDED OFF TO REVEAL BASE PRIME |
|---|---|---|---|---|---|
| 2 | 3 | +1 | N/A[4] | N/A | 3 |
| 3 | 5 | +2 | 1.6180 | 4.6180 | 5 |
| 5 | 7 | +2 | 2.6180 | 7.236 | 7 |
| 7 | 11 | +4 | 3.6180 | 10.854 | 11 |
| 11 | 13 | +2 | 1.6180 | 12.472 | 13 |
| 13 | 17 | +4 | 4.6180 | 17.09 | 17 |
| 17 | 19 | +2 | 1.6180 | 18.708 | 19 |
| 19 | 23 | +4 | 3.6180 | 23.326 | 23 |
| 23 | 29 | +6 | 5.6180 | 28.994 | 29 |
| 29 | 31 | +2 | 1.6180 | 30.562 | 31 |
| 31 | 37 | +6 | 6.6180 | 37.18 | 37 |
| 37 | 41 | +4 | 3.6180 | 40.798 | 41 |
| 41 | 43 | +2 | 2.6180 | 43.416 | 43 |
| 43 | 47 | +4 | 3.6180 | 47.034 | 47 |
| 47 | 53 | +6 | 5.6180 | 52.652 | 53 |
| 53 | 59 | +6 | 6.6180 | 59.27 | 59 |
| 59 | 61 | +2 | 1.6180 | 60.888 | 61 |
| 61 | 67 | +6 | 5.6180 | 66.506 | 67 |
| 67 | 71 | +4 | 4.6180 | 71.124 | 71 |
| 71 | 73 | +2 | 1.6180 | 72.742 | 73 |
| 73 | 79 | +6 | 6.6180 | 79.36 | 79 |
| 79 | 83 | +4 | 3.6180 | 82.978 | 83 |
| 83 | 89 | +6 | 5.6180 | 88.596 | 89 |
| 89 | 97 | +8 | 8.6180 | 97.214 | 97 |
| 97 | 101 | +4 | 3.6180 | 100.832 | 101 |
| 101 | 103 | +2 | 2.6180 | 103.45 | 103 |
| 103 | 107 | +4 | 3.6180 | 107.068 | 107 |
| 107 | 109 | +2 | 1.6180 | 108.686 | 109 |
| 109 | 113 | +4 | 4.6180 | 113.304 | 113 |
| 113 | 127 | +13 | 13.6180 | 126.922 | 127 |

---

3 Actually, the reciprocal of Phi, 0.6180, is the base number to use in this procedure.

4 "One" is the increment between the only even prime, 2, and three. There is no Phi number to assess in this case.

| Natural Numbers + 0.6180 | Rounded off | PRIME | 1/7 DIGITS (Sum Mod 9) | SUPERNAL TRIAD |
|---|---|---|---|---|
| (1) | (1) | (1) | (1) | |
| 2.6180 | 3 | 2 | 2 | |
| 3.6180 | 4 | 3 | 3 | |
| 4.6180 | 5 | 5 | 5 | |
| 5.6180 | 6 | 7 | 7 | |
| 6.6180 | 7 | 11 | 2 | |
| 7.6180 | 8 | 13 | 4 | |
| 8.6180 | 9 | 17 | 8 | |
| 9.6180 | 10 | 19 | 1 | |
| 10.6180 | 11 | 23 | 5 | |
| 11.6180 | 12 | 29 | 2 | |
| 12.6180 | 13 | 31 | 4 | |
| 13.6180 | 14 | 37 | 1 | |
| 14.6180 | 15 | 41 | 5 | |
| 15.6180 | 16 | 43 | 7 | |
| 16.6180 | 17 | 47 | 2 | |
| 17.6180 | 18 | 53 | 8 | |
| 18.6180 | 19 | 59 | 5 | |
| 19.6180 | 20 | 61 | 7 | |
| 20.6180 | 21 | 67 | 4 | |
| 21.6180 | 22 | 71 | 8 | |
| 22.6180 | 23 | 73 | 1 | |
| 23.6180 | 24 | 79 | 7 | |
| 24.6180 | 25 | 83 | 2 | |
| 25.6180 | 26 | 89 | 8 | |
| 26.6180 | 27 | 97 | 7 | |
| 27.6180 | 28 | 101 | 2 | |
| 28.6180 | 29 | 103 | 4 | |
| 29.6180 | 30 | 107 | 8 | |
| 30.6180 | 31 | 109 | 1 | |
| 31.6180 | 32 | 113 | 5 | |
| 32.6180 | 33 | 127 | 1 | |

Based on the first two columns, the reciprocal of Phi, 0.6180, generates the natural numbers. and could be considered quasi-increments (secondary) of the natural numbers.

# UNIMATH ASSSESSMENT OF PRIME NUMBERS

Preliminary ideas:

- All numbers and number sequences can be reducedd to positive and negative binary pairs (binarics); e.g. +7 -7.

- It has been found tha a binaric can be assigned a value connected with Phi and the reciprocal of phi and paired with a prime number binaric.

- As understood in number theory: $\phi - \dfrac{1}{\phi} = 1$

- When you assign the positive number of a binaric to phi and the negative component to the reciprocal of phi, the result is always the square root of phi; 2.23606….

- It is always true that when you assign these values to consecutive prime numbers, the value always comes to negative primes, e.g.., -1, -3, -5, -7; when these numbers of the prime increments are added to the negative number, the result is always 1 (one).

- <u>One</u> of the formulas relating the prime numbers to phi is:

$$P_1 + \phi - P_2 + \dfrac{1}{\phi} = \phi V_b$$

*..which reads Prime 1 plus Phi minus Prime 2 + the reciprocal of phi equals the intrinstic Binaric Phi value.*

- Conjecture: Phi and the reciprocal of phi represents the binaric fractal of numbers.

- Conjecture: The prime numbers increase in a pattern of a logarithmic spiralar wave form.

$$P_1 + \phi - P_2 + \frac{1}{\phi} = \phi V_b$$

The above equation describes the generation of Phi values. These values play a part in the behavior of prime numbers.

*The equation reads:  P1 (the first prime in a binaric) plus Phi (1.61803399...) minus P2 (the second prime in a binaric) plus the reciprocal of Phi, 0.61803399...) results in the BINARIC Phi Value.*

The first step is to assess consecutive prime binarics is to determine the values in the above equation.

**Example 1:**  We will start with finding the difference between  **3 (P1) and 5 (P2):**

| 3(P1) | 5(P2) |
|---|---|
| 1.6180 | 0.6180 |
| +3. 000 | +5.0000 |
| 4.6180 | 5.6180 |

*Subtract 5.6180 from 4.6180 to get a BINARIC Phi Value of* -1

---

5 Binaric is the name given to a number binary; a pair of numbers...

**Example 2:**

<u>5(P1)</u>                                                  <u>7(P2)</u>

$$
\begin{array}{ll}
\quad 1.6180 & \quad 0.6180 \\
\underline{+5.0000} & \underline{+\,7.000} \\
\quad 6.6180 & \quad 7.6180
\end{array}
$$

*Subtract 7.6180 from 6.6180 for a BINARIC Phi Value of* **-1**

**Example 3:**

<u>7 (P1)</u>                                                 <u>11(P2)</u>

$$
\begin{array}{ll}
\quad 1.6180 & \quad 0.6180 \\
\underline{+\,7.0000} & \underline{+11.0000} \\
\quad 8.6180 & \quad 11.6180
\end{array}
$$

**Subtract 11. 6180 from 9.2360 and the BINARIC Phi Value is** **-3**

# PRIME BINARICS AND PHI RELATIONSHIP CHART

| PRIME 1 | PRIME 2 | INCREMENT | BINARIC $\Phi$ VALUE | $\Sigma$ $P_1 + P_2$ $\Phi$ VALUE |
|---|---|---|---|---|
| 2 | 3 | 1 | 0 | 1 |
| 3 | 5 | 2 | -1 | 1 |
| 5 | 7 | 2 | -1 | 1 |
| 7 | 11 | 4 | -3 | 1 |
| 11 | 13 | 2 | -1 | 1 |
| 13 | 17 | 4 | -3 | 1 |
| 17 | 19 | 2 | -1 | 1 |
| 19 | 23 | 4 | -3 | 1 |
| 23 | 29 | 6 | -5 | 1 |
| 29 | 31 | 2 | -1 | 1 |
| 31 | 37 | 6 | -5 | 1 |
| 37 | 41 | 4 | -3 | 1 |
| 41 | 43 | 2 | -1 | 1 |
| 43 | 47 | 4 | -3 | 1 |
| 47 | 53 | 6 | -5 | 1 |
| 53 | 59 | 6 | -5 | 1 |
| 59 | 61 | 2 | -1 | 1 |
| 61 | 67 | 6 | -5 | 1 |
| 67 | 71 | 4 | -3 | 1 |
| 71 | 73 | 2 | -1 | 1 |
| 73 | 79 | 6 | -5 | 1 |
| 79 | 83 | 4 | -3 | 1 |
| 83 | 89 | 6 | -5 | 1 |
| 89 | 97 | 8 | -7 | 1 |
| 97 | 101 | 4 | -3 | 1 |

# MORE PRIME PATTERNS

## THE FIRST 103 PRIME NUMBERS

| 2 | 3 | 5 | 7 | 11 | 13 | 17 | 19 | 23 | 29 |
|---|---|---|---|---|---|---|---|---|---|
| 31 | 37 | 41 | 43 | 47 | 53 | 59 | 61 | 67 | 71 |
| 73 | 79 | 83 | 89 | 97 | 101 | 103 | 107 | 109 | 113 |
| 127 | 131 | 137 | 139 | 149 | 151 | 157 | 163 | 167 | 173 |
| 179 | 181 | 191 | 193 | 197 | 199 | 211 | 223 | 227 | 229 |
| 233 | 239 | 241 | 251 | 257 | 263 | 269 | 271 | 277 | 281 |
| 283 | 293 | 307 | 311 | 313 | 317 | 331 | 337 | 347 | 349 |
| 353 | 359 | 367 | 373 | 379 | 383 | 389 | 397 | 401 | 409 |
| 419 | 421 | 431 | 433 | 439 | 443 | 449 | 457 | 461 | 463 |
| 467 | 479 | 487 | 491 | 499 | 503 | 509 | 521 | 523 | 541 |

| 547 | 557 | 563 | | | | | | | |
|---|---|---|---|---|---|---|---|---|---|

# THE FIRST 103 PRIME INCREMENTS CONTIGUOUSLY ASSESSED[6]

+1 +2 +2 -6 0 0 +2 -2 +6 -6 +8 -7 +1 -1 +7 -6 -2 +2 +4 -3 -3 +3 -1

+1 +3 -2 -2 +2 +4 -3 -5 +5 +1 0 -6 +6 -4 +4 +2 -1 -5 +5 +1 0 -2 -6 -1

+1 0 -1 +3 -3 -1 +7 -6 -1 +9 -8 0 +2 -2 +1 +5 -6 +2 -2 0 +2 +4 -6

+2 +6 -8 +3 +5 -8 +4 -4 0 +4 +2 -6 +5 -3 -2 +5 +1 -6 +6 -4 -2 +6 +2

-8 +7 -7 0 +8 -8 0 +8 -6 -2 +8 -2 -6 +8 0 -7 -1 0 +1 0 +1 -1 0 +5 -5

0 +7 -7 +1 0 -1 +1 +6 -7 +2 -3 +1 +3 -4 +1+3 +2 -5 +4 -3 -1 +4 +3 -7

+5 -6 +1 +5 0 -5 +6 -7 +1 +6 -5 -1 +7 -6 0 -3 +7 -4 -2 0 +2 -2 +2 0 -

2 +6 -4 0 -2 +2 0 +4 -4 +1 +3 -4 +1 +5 -6 +2 -2 0 +2 +4 -6 +3 +1 -4

+4 -4 0 +4 +2 -6 +5 -5 0 +5 +1 -6 +6 -2 -3 -4 +1 +3 -4 +9 -5 -3 +8 -5

-2 -1 +3 -1 -2 +3 -1 0 +1 -1 +6 -5 0 -1 +1 0 +5 -5 +1 +2 -3 +2 -5 +3

+2 -3 +1 +2 +1 -3 +3 +2 -5 +4 -1 -3 +5 -8 +3 +5 0 -4 -5 +3 +2 -5 +9

-4 -3 -1 +4 -3 +1 +2 -1 -3 +4 -1 +3 -2 0 +2 -2 +1 -3  **=0**

---

6 The sum of the increments of the first 103 primes is 0.

# HOMOGENEOUS BINARIC PATTERNS

+1 +2 +2 -6 0 0 +2 -2 +6 -6 +8 -7 +1 -1 +7 -6 -2 +2

+4 -3 -3 +3 -1 +1 +3 -2 -2 +2 +4 -3 -5 +5 +1 0 -6 +6 -4

+4 +2 -1 -5 +5 +1 0 -2 -6 -1 +1 0 -1 +3 -3 -1 +7 -6 -1

+9 -8 0 +2 -2 +1 +5 -6 +2 -2 0 +2 +4 -6 +2 +6 -8 +3

+5 -8 +4 -4 0 +4 +2 -6 +5 -3 -2 +5 +1 -6 +6 -4 -2 +6

+2 -8 +7 -7 0 +8 -8 0 +8 -6 -2 +8 -2 -6 +8 0 -7 -1 0

+1 0 +1 -1 0 +5 -5 0 +7 -7 +1 0 -1 +1 +6 -7 +2 -3 +1

+3 -4 +1+3 +2 -5 +4 -3 -1 +4 +3 -7 +5 -6 +1 +5 0 -5

+6 -7 +1 +6 -5 -1 +7 -6 0 -3 +7 -4 -2 0 +2 -2 +2 0 -2

+6 -4 0 -2 +2 0 +4 -4 +1 +3 -4 +1 +5 -6 +2 -2 0 +2 +4

-6 +3 +1 -4 +4 -4 0 +4 +2 -6 +5 -5 0 +5 +1 -6 +6 -2 -3

-4 +1 +3 -4 +9 -5 -3 +8 -5 -2 -1 +3 -1 -2 +3 -1 0 +1 -1

+6 -5 0 -1 +1 0 +5 -5 +1 +2 -3 +2 -5 +3 +2 -3 +1 +2

+1 -3 +3 +2 -5 +4 -1 -3 +5 -8 +3 +5 0 -4 -5 +3 +2 -5

+9 -4 -3 -1 +4 -3 +1 +2 -1 -3 +4 -1 +3 -2 0 +2 -2 +1 -3

# HOMOGENOUS & NON-HOMOGENOUS BINARIC PATTERNS

+1 +2 +2 -6 0 0 +2 -2 +6 -6 +8 -7 +1 -1 +7 -6 -2 +2

+4 -3 -3 +3 -1 +1 +3 -2 -2 +2 +4 -3 -5 +5 +1 0 -6 +6 -4

+4 +2 -1 -5 +5 +1 0 -2 -6 -1 +1 0 -1 +3 -3 -1 +7 -6 -1

+9 -8 0 +2 -2 +1 +5 -6 +2 -2 0 +2 +4 -6 +2 +6 -8 +3

+5 -8 +4 -4 0 +4 +2 -6 +5 -3 -2 +5 +1 -6 +6 -4 -2 +6

+2 -8 +7 -7 0 +8 -8 0 +8 -6 -2 +8 -2 -6 +8 0 -7 -1 0

+1 0 +1 -1 0 +5 -5 0 +7 -7 +1 0 -1 +1 +6 -7 +2 -3 +1

+3 -4 +1 +3 +2 -5 +4 -3 -1 +4 +3 -7 +5 -6 +1 +5 0 -5

+6 -7 +1 +6 -5 -1 +7 -6 0 -3 +7 -4 -2 0 +2 -2 +2 0 -2

+6 -4 0 -2 +2 0 +4 -4 +1 +3 -4 +1 +5 -6 +2 -2 0 +2 +4

-6 +3 +1 -4 +4 -4 0 +4 +2 -6 +5 -5 0 +5 +1 -6 +6 -2 -3

-4 +1 +3 -4 +9 -5 -3 +8 -5 -2 -1 +3 -1 -2 +3 -1 0 +1 -1

+6 -5 0 -1 +1 0 +5 -5 +1 +2 -3 +2 -5 +3 +2 -3 +1 +2

+1 -3 +3 +2 -5 +4 -1 -3 +5 -8 +3 +5 0 -4 -5 +3 +2 -5

+9 -4 -3 -1 +4 -3 +1 +2 -1 -3 +4 -1 +3 -2 0 +2 -2 +1 -3

+1 +2 +2 -6 0 0 +2 -2 +6 -6 +8 -7 +1 -1 +7 -6 -2 +2

+4 -3 -3 +3 -1 +1 +3 -2 -2 +2 +4 -3 -5 +5 +1 0 -6 +6 -4

+4 +2 -1 -5 +5 +1 0 -2 -6 -1 +1 0 -1 +3 -3 -1 +7 -6 -1

+9 -8 0 +2 -2 +1 +5 -6 +2 -2 0 +2 +4 -6 +2 +6 -8 +3

+5 -8 +4 -4 0 +4 +2 -6 +5 -3 -2 +5 +1 -6 +6 -4 -2 +6

+2 -8 +7 -7 0 +8 -8 0 +8 -6 -2 +8 -2 -6 +8 0 -7 -1 0

+1 0 +1 -1 0 +5 -5 0 +7 -7 +1 0 -1 +1 +6 -7 +2 -3 +1

+3 -4 +1+3 +2 -5 +4 -3 -1 +4 +3 -7 +5 -6 +1 +5 0 -5

+6 -7 +1 +6 -5 -1 +7 -6 0 -3 +7 -4 -2 0 +2 -2 +2 0 -2

+6 -4 0 -2 +2 0 +4 -4 +1 +3 -4 +1 +5 -6 +2 -2 0 +2 +4

-6 +3 +1 -4 +4 -4 0 +4 +2 -6 +5 -5 0 +5 +1 -6 +6 -2 -3

-4 +1 +3 -4 +9 -5 -3 +8 -5 -2 -1 +3 -1 -2 +3 -1 0 +1 -1

+6 -5 0 -1 +1 0 +5 -5 +1 +2 -3 +2 -5 +3 +2 -3 +1 +2

-3 +3 +2 -5 +4 -1 -3 +5 -8 +3 +5 0 -4 -5 +3 +2 -5 +9

-4 -3 -1 +4 -3 +1 +2 -1 -3 +4 -1 +3 -2 0 +2 -2 +1 -3

---

7 Reciprocals of Digits of 7 = 65%; 3-6-9= 24%; and 0=11%

## INCIDENCES OF NUMBERS IN 103 PRIMES CONTIGUOUSLY ASSESSED

| NUMBER | OCCURENCE |
| --- | --- |
| 1 | 49 |
| 2 | 49 |
| 3 | 35 |
| 4 | 29 |
| 5 | 30 |
| 6 | 29 |
| 7 | 13 |
| 8 | 12 |
| 9 | 3 |
| 0 | 30 |

## DEDUCTION

Prime Conjecture: The limit of prime increments as n approaches infinity equals Zero (0).

$$\lim_{n \to \infty} \Delta_p = 0$$

# ADDITIVE ASSESSMENT

The additive assessment of a number sequence is the process of sampling bits of information in larger and larger pieces and assessing the samples. For example, take a number sequence and analyze the first two numbers of the sequence, and then the first three, the first four, etc. until the entire sequence has been assessed.  In increment analysis, additive assessment is used to assess the increments of larger and larger blocks of a number sequence.  Zeroes are added before and after the segment. When this is done, the segment always sums to zero, with a graduated pattern of positive and negative numbers.  Following, the additive assessment technique is demonstrated for the number *e* (the basis of natural logarithms) up to the 25$^{th}$ integer, **pi**,  primes (up to the 13$^{th}$ prime) the first **10 Fibonacci numbers**, the **reciprocal of the number seven**, **the reciprocal of number thirteen** and **phi.**

# PRIMES

**(The first thirteen)**

*2  3  5  7  11  13  17  19  23  29  31  37  41*

| | |
|---|---|
| 020 = | +2  - 2 |
| 0230 = | +3  - 3 |
| 02350 = | +5  - 5 |
| 023570 = | +7  - 7 |
| 02357 110 = | +11 -11 |
| 02357 11  130 = | +13 -13 |
| 02357 11 13 170 = | +17 -17 |
| 02357 11 13 17  190 = | +19 -19 |
| 02357 11 13  17 19  230 = | +23 - 23 |
| 02357 11  13  17  19 23  290 = | +29 -29 |
| 02357 11  13 17  19 23 29  310 = | +31 -31 |
| 02357 11  13 17  19 23 29 31  370 = | +37 -37 |
| 02357 11  13 17  19 23 29 31 37  410 = | +41 -41 |

**(PHI)  (THE GOLDEN MEAN) – 1.6180339…**

| | | | |
|---|---|---|---|
| 010 | | +1 | -1 |
| 0160 | | +6 | -6 |
| 01610 | | +6 | -6 |
| 016180 | | +13 | -13 |
| 0161800 | | +13 | -13 |
| 01618030 | | +16 | -16 |
| 016180330 | | +16 | -16 |
| 0161803390 | | +22 | -22 |

The prime and Fibonacci numbers "say their own names."   In other words, the last prime or Fibonacci of each segment gives the name of the segment, which follows the natural order of each sequence with no repetition characteristic of the other numbers assessed thus far.  This might be true for every uniformly monotonic increasing number sequence.

## THE FIBONACCI NUMBERS

Fibonacci numbers were first demonstrated by Leonardo of Pisa, better known as Fibonacci, in his book titled **Liber Abaci** published in 1202. In it he discussed the number of rabbits that would be born from a pair of rabbits at the beginning of every month.  The resulting number of rabbits revealed the following pattern: 1,1,2,3,5,8,13,21,34,55, etc.  This series converges to a number called the "golden ratio," which is rounded off here to **1.6180339887**….. Basically, you take a term, add it to the previous term, and the number converges to the golden ratio.  This number is found in many places in nature.  The formula that generates this number is:

$$\frac{(1 + \sqrt{5})}{2} = 1.6180339887\ldots..$$

In an examination of the first 500 Fibonacci numbers, Increment Analysis has revealed an interesting pattern– when the digits are summed modulo nine, a recurring sequence with a mirror pattern emerges. The sequence is:

**112358437189887641562819…1…**…..

**This series of numbers repeats continually beginning again at every 24th digit.** The **24th digit always sums to nine**.  The increment pattern that emerges is:

**0 +1  +1  +2  +3  -4  -1  +4  -6  +7  +1  -1**

**0 −1  -1   -2  -3  +4  +1  -4  +6  -7  +8  -8**

*The series sums to a perfect mirror with an anomaly, +1 +8, -1 –8, which sums to +9 and –9, which sums to zero.8*

---

8 The Lucas numbers are similar to the Fibonacci numbers, but they start with 2 instead of 1, and when the digits are summed modulo nine, a recurring sequence emerges that also repeats at every 24th number.

# ADDITIVE SIGNATURES

As demonstrated in the previous examples, every number sequence has an additive signature that is characteristic of its assessment. The signatures of the numbers thus far mentioned are shown below.

| | |
|---|---|
| **e** | ±<br><br>2,7,7,14,14,20,20,27,27,33,34,38,38,38,42,43,43,44,46,46,49,49,51,57,57 |
| **Primes** | ±<br><br>2,3,5,7,11,13,17,19,23,29,31,37,41 |
| **π** | ±<br><br>3,3,6,6,10,14,14,18,18,18,20,23,24,24,26,26,26,27,32,32,34,34,38,38,38,38,38,43,45,45,45 |
| **Fibona-cci #** | ±<br><br>1,1,2,3,5,8,13,21,34,55 |
| **1/7** | ±<br><br>1,4,4,10,10,12,12,15,15,21,21,23,23,26,26,32,34,34,34 |
| **1/13** | ±<br><br>7,7,10,10,11,11,18,18,21,21,22,22,29 |
| **Phi** | ±<br><br>1,6,6,13,13,16,16,22 |

# MISCELLANEOUS PRIME INFORMATION

What has been revealed by Increment Analysis is:

- No prime number with two or more digits sums to *6, 3, or 9.*

- No prime with two or more digits ends with the number *5.*

- The final digit of any prime number of two or more digits *can only be 1, 3, 7, or 9.*

- ***All primes when viewed Sum Modulo 9 reduce to digits of the reciprocal of the number 7.***

- There is a yin yang configuration of the primes.  The only even prime, number 2, is separated from every other prime by odd numbered increments. Every odd prime is separated from every other prime by an even numbered increment.

- Though it has been shown that the frequency of twin primes decreases as one gets to larger numbers, Increment Analysis demonstrates that the frequency of larger increments increases as one gets to larger numbers.

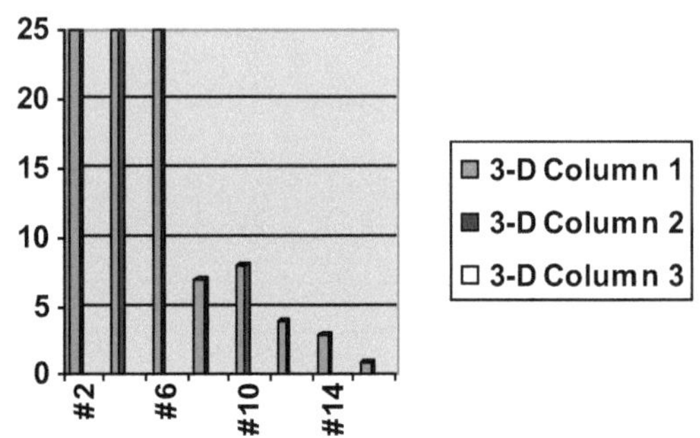


- ♦ # 2 appears 25 x

- ♦ # 4 appears 25 x

- ♦ # 6 appears 25 x

- ♦ # 8  appears 7 x

- ♦ # 10 appears 8 x

- ♦ # 12 appears 4 x

- ♦ # 14 appears 3 x

- ♦ *# 18 appears once*

*In addition, the increment between the* **20,861$^{st}$ and 20,862$^{nd}$ prime** (235,397 and

235,439) is **42.**

**THE FIRST 101 PRIMES SUMMED MODUL0 9 DEMONSTRATING DIGITS FROM
THE RECIPROCAL OF THE NUMBER SEVEN**

**(1)23 = The Supernal Triad**

**572481524157285748128724815152457415281248147248578528172451572745 7287**

**415215457817287248215485 8117**

**PRIME EXCLUSION PRINCIPLES**

Based on the foregoing, as previously pointed out, it is safe to assume that, when

assessing a number with two or more digits, you know it ***is not prime*** if the final digit is a 5. ***It is also not prime if the sum modulo nine is not one of the digits of the reciprocal of seven.*** In other words, **there are no 3s, 6s, or 9s** when assessed sum modulo 9.

## PRIMAL ELEMENTS

When considering the Periodic Table of the Elements, the prime number elements are arranged in interesting patterns. The periodic table is divided into classes numbered from 1 through 18. The $18^{th}$ division is comprised of the "noble gases."

| I | II | IIIb | IVb | Vb | VIb | VIIb | VIIIb | | | Ib | IIb | III | IV | V | VI | VII | 0 |
|---|----|------|-----|----|-----|------|-------|--|--|----|-----|-----|----|---|----|-----|---|
| 1 | 2 | 3 | 4 | 5 | 6 | 7 | 8 | 9 | 10 | 11 | 12 | 13 | 14 | 15 | 16 | 17 | 18 |
| Li | none | Ac | none | V | none | Tc | none | Mt | none | Cu | none | B | none | N | none | Cl | He |
| Na | | | | Nb | | Bh | | | | Ag | | Al | | Bi | | I | |
| K | | | | Ta | | Pm | | | | Au | | Ga | | Md | | Lu | |
| Rb | | | | Pr | | | | | | Bk | | Ho | | | | Lr | |

Totals

| 4 | 0 | 1 | 0 | 4 | 0 | 3 | 0 | 1 | 0 | 4 | 0 | 4 | 0 | 3 | 0 | 4 | 1 |
|---|---|---|---|---|---|---|---|---|---|---|---|---|---|---|---|---|---|

The "prime" elements, i.e., the elements with the same number of protons as the prime numbers, tend to alternate, and they tend to fall in prime numbered columns, with a few exceptions. When the total number of prime elements in each group is reduced to a number, the following sequence results:

**4  0  1  0  4  0  3  0  1  0  4  0  4  0  3  0  4  1;**

**401040301040403041 =**

**040104030104040304 10 =**

**+4 -4 +1 -1 +4 -4 +3 -3 +1 -1 +4 -4 +4 -4 +3 -3 +4 -3 -1 =**

**+28  -28;**

**Sum modulo 9 =**

**+10   -10 =**

**+1      -1**

***This last section, the analysis of the resulting number sequence reveals a pattern of primes in the natural elements that resolves to the Binaric +1 -1.***

# SQUARE BUSINESS

In this section I will share an examination of the increments in "squares"-- zero sum squares, contiguous prime triads, no-contiguous prime triads, and non-prime triads.

## *2 x 2 SQUARES*

I have developed a class of 2 x 2 squares that utilizes zero and a negative number to form Zero Sum squares. Zero Sum squares generate series of positive and negative numbers that sum to zero.  The following ***is a Zero Sum 2 x 2 square*** comprised of the first three primes and zero wherein each row, column or diagonal sums to positive and negative prime numbers that together sum to zero. I have also developed non-zero 2 x 2 squares that generate the same numbers when the horizontal, vertical, and diagonal aspects are assessed.

*METHODOLOGY*: in the 2 x 2 squares the numbers are added clockwise with the lowest number placed in the bottom left square; the next highest number placed in the square above it: Zero is placed in the top right square, and a negative number is placed in the bottom right square. Next, the horizontal squares are added together, the vertical columns are added together, and the diagonals are added together. The result is that the horizontals, verticals, and diagonal numbers add to the same number. ***In the case of Zero Sum Squares, the initial square, the bottom left, is always +2.***

# +10 -10 Zero Sum Prime Number square

<table>
<tr><td>3</td><td>0</td></tr>
<tr><td>2</td><td>-5</td></tr>
</table>

Horizontals:   +3  -3

Verticals:   +5 -5

Diagonals:   +2 -2

# +14 -14 Zero Sum Prime Number square

| | |
|---|---|
| 5 | 0 |
| 2 | -7 |

Horizontals:   +5 -5

Verticals:   +7 -7

Diagonals:   +2 -2

# +38 -38   Zero Sum Prime Number square

<table>
<tr><td>17</td><td>0</td></tr>
<tr><td>2</td><td>-19</td></tr>
</table>

**Horizontals:**   +17  -17

**Verticals:**   +19  -19

**Diagonals:**   +2   -2

# +62 -62 Zero Sum Prime Number Square

| | |
|---|---|
| 29 | 0 |
| 2 | -31 |

Horizontals: +29 -29

Verticals: +31 -31

Diagonals: +2 -2

# 2 X 2 Contiguous Prime Square Triads

## 2-3-5

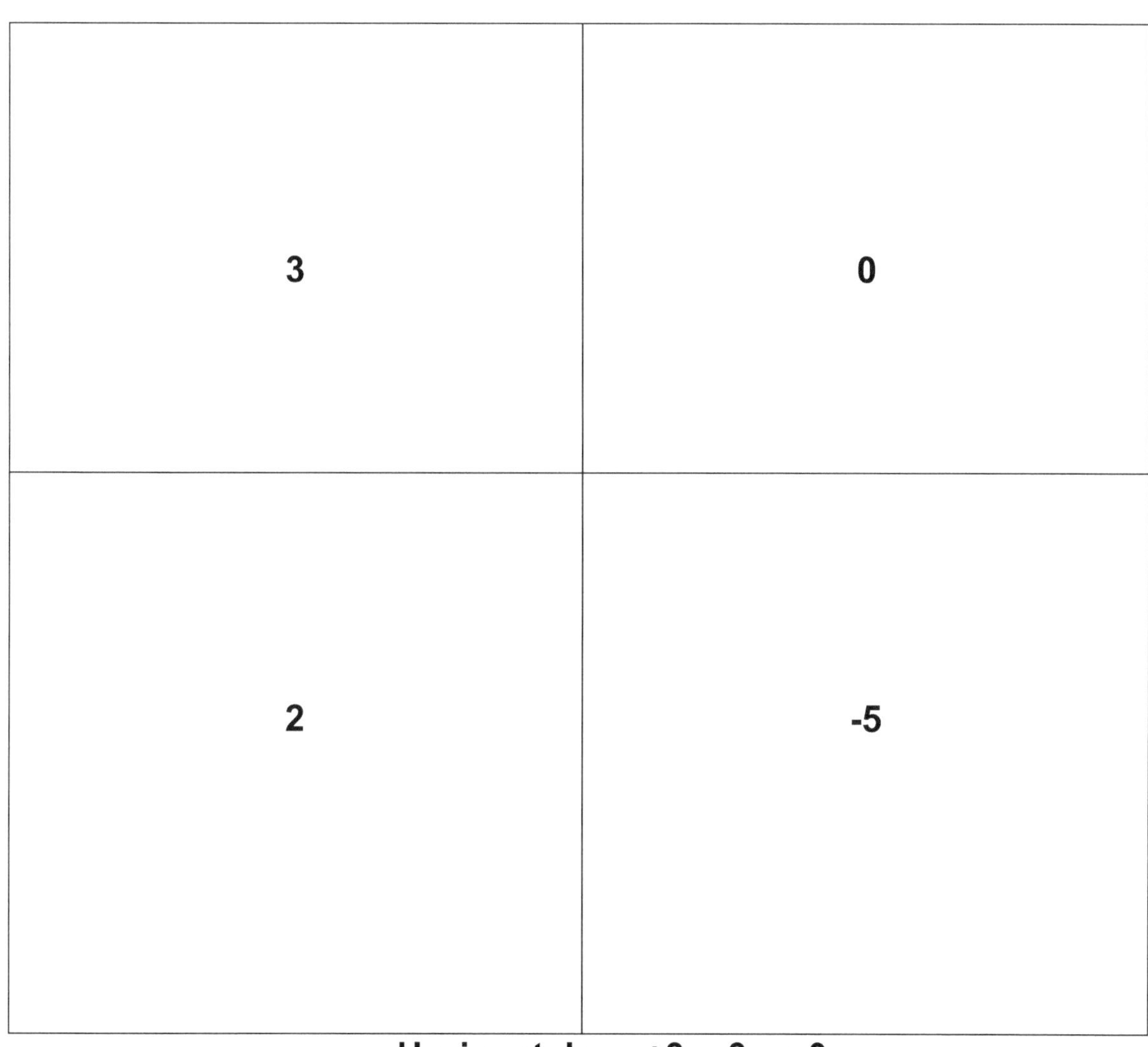



Horizontals    +3  -3 = 0

Verticals    +5  -5 = 0

Diagonals    +2  -2 = 0

SUM = 0

# 5-7-11 Triad

<table>
<tr><td>7</td><td>0</td></tr>
<tr><td>5</td><td>-11</td></tr>
</table>

Horizontals    +7    -6 = +1

Verticals    +12   -11 = +1

Diagonals    +5   -4 = +1

SUM = 3

# 11-13-17 Triad

<table>
<tr><td>13</td><td>0</td></tr>
<tr><td>11</td><td>-17</td></tr>
</table>

Horizontals:  +13     -6 =  +7

Verticals:       +24   -17 = +7

Diagonals:     +11     -4 = +7

SUM = 21

# 17 – 19 - 23 Triad

<table>
<tr><td>19</td><td>0</td></tr>
<tr><td>17</td><td>-23</td></tr>
</table>

Horizontals +19   -6   =   13

Verticals      +36 -23 =   13

Diagonals     +17 -4 =    13

SUM = 39

# 23 – 29 - 31 Triad

<table>
<tr><td>29</td><td>0</td></tr>
<tr><td>23</td><td>-31</td></tr>
</table>

Horizontals +29   -8 = 21

Verticals    +52   -31 = 21

Diagonals    +23   -2 = 21

SUM = 63

# 31– 37– 41 Triad

<table>
<tr><td>37</td><td>0</td></tr>
<tr><td>31</td><td>-41</td></tr>
</table>

Horizontals +37 -10 = 27

Verticals    +68 -41= 27

Diagonals    +31 -4 = 27

SUM = 81

# 41-43-47 Triad

<table>
<tr><td>43</td><td>0</td></tr>
<tr><td>41</td><td>-47</td></tr>
</table>

Horizontals +43 -6 = 37

Verticals    +84 -47 = 37

Diagonals    +41 -4 = 37

SUM = 111

# 47-53-59 Triad

<table>
<tr><td>53</td><td>0</td></tr>
<tr><td>47</td><td>-59</td></tr>
</table>

Horizontals  +53  -12 = 41

Verticals     +100 -59 = 41

Diagonals    +47  -6 = 41

SUM = 123

# 59 – 61 - 67 Triad

<table>
<tr><td>61</td><td>0</td></tr>
<tr><td>59</td><td>-67</td></tr>
</table>

**Horizontals +61 -8 = 53**

**Verticals   +120 -67 = 53**

**Diagonals   +59  -6 = 53**

**SUM = 159**

# 67-71-73 Triad

<table>
<tr><td>71</td><td>0</td></tr>
<tr><td>67</td><td>-73</td></tr>
</table>

Horizontals +71 -6 = 65

Verticals +138 -73 = 65

Diagonals   +67 -2 = 65

SUM = 195

# 73 – 79 - 83 Triad

<table>
<tr><td>79</td><td>0</td></tr>
<tr><td>73</td><td>-83</td></tr>
</table>

Horizontals +79 -10 = 69

Verticals + 152 -83 = 69

Diagonals +73 -4 = 69

SUM = 207

# 83 – 89 – 97 Triad

<table>
<tr><td>89</td><td>0</td></tr>
<tr><td>83</td><td>-97</td></tr>
</table>

Horizontals +89 -14 = 75

Verticals +172 -97 = 75

Diagonals +83 -8 = 75

SUM = 225

# 97-101 - 103 Triad

|  |  |
|---|---|
| 101 | 0 |
| 97 | -103 |

Horizontals +101 -6 = 95

Verticals +198 -103 = 95

Diagonals +97 -2    = 95

SUM: 285

# 103 -107 - 109 Triad

| | |
|---|---|
| 107 | 0 |
| 103 | -109 |

Horizontals +107 -6 = 101

Verticals +210 -109 = 101

Diagonals +103 – 2 = 101

SUM: 303

*The values of 2 x 2 Prime squares always result in numbers that resolve to 3, 6, or 9 when assessed sum modulo 9.*

| PRIME TRIAD | SQUARE VALUE | SUM MOD 9 |
|---|---|---|
| 2-3-5 | 0 | 0 |
| 5-7-11 | 3 | 3 |
| 11-13-17 | 21 | 3 |
| 17-19-23 | 39 | 3 |
| 23-29-31 | 63 | 9 |
| 31-37-41 | 81 | 9 |
| 41-43-47 | 111 | 3 |
| 47-53-59 | 123 | 6 |
| 59-61-67 | 159 | 6 |
| 67-71-73 | 195 | 6 |
| 73-79-83 | 207 | 9 |
| 83-89-97 | 225 | 9 |
| 97-101-103 | 285 | 6 |
| 103-107-109 | 303 | 6 |
| **TOTALS** | **1,809** | **9** |

This is the summary chart connected with the previous prime triads.

# NON-CONTIGUOUS PRIME TRIADS

## 7–11-13 TRIAD

<table>
<tr><td>11</td><td>0</td></tr>
<tr><td>7</td><td>-13</td></tr>
</table>

**Horizontals +11 -6 = 5**

**Verticals +18 -13 = 5**

**Diagonals +7 -2 = 5**

**SUM: 15 = 6**

# 13-29-83 Triad

<table>
<tr><td>29</td><td>0</td></tr>
<tr><td>13</td><td>-83</td></tr>
</table>

**Horizontals +29 -70 = -41**

**Verticals +42 -83 = -41**

**Diagonals 13 -54= -41**

**SUM: -123**

# 43-73-101 Triad

<table>
<tr><td>*73*</td><td>*0*</td></tr>
<tr><td>*43*</td><td>*-101*</td></tr>
</table>

**Horizontals +73 -58 = 15**

**Verticals +116 -101 = 15**

**Diagonals +43 -28 = 15**

**SUM = 45 = 9**

# Non-Prime Triad

## 28-55-94

<table>
<tr><td>55</td><td>0</td></tr>
<tr><td>28</td><td>-94</td></tr>
</table>

**Horizontals +55 -66 =-11**

**Verticals +83 –94 =  -11**

**Diagonals +28 -39 = -11**

**SUM: -33**

# CONCLUSIONS AND IMPLICATIONS

Based on the foregoing, it can be safely conjectured that the building blocks of the prime numbers are the digits of phi. There is a direct correlation between the increments of contiguous prime numbers and the iterations of phi; i.e., the increment is added to the reciprocal of phi ($\Delta$ or $\Delta$ -1). For example, if the increment between contiguous prime numbers is 8, the Phi value added to that number to generate the next prime is **8**.6180. Similarly, if the increment is 4, the number **4**.6180 must be added to the previous prime to generate the new one. For some of the numbers, the value added will be 1 (and only 1) less than the increment.

Also, it has been found that the Fibonacci numbers, which converge to Phi, have a 24 digit cycle when the numbers are reduced to a single digit (sum mod 9), with a culminating +9 -9 binaric. The sequence is cyclic and repeats continuously. This indicates that there is possibly a circular, aspect of the prime numbers. Moreover, the prime numbers always reduce sum mod 9 to digits of the reciprocal of seven (7), which is a circular number that repeats continually and exhibits a

**+11 -11** binaric. In addition, 2x2 squares reveal interesting patterns in the prime numbers, and Phi values added to numbers result in unity. Based on these findings, the prime numbers are organized around iterations of Phi, and they may have a circular component.

Mathematician Gregory J. Chaitin of the IBM Thomas J. Watson Research Center, author of *The Limits of Mathematics* suggests that the structure of arithmetic is random. He states, "Although almost all numbers are random, there is no formal axiomatic system that will allow us to prove this fact." Unimath seriously challenges this idea.

To summarize, all number sequences, have increments that sum to zero and result in positive and negative binarics. If you add zeros before and after any sequence the result after assessing increments is zero.  All numbers placed in a circle sum to zero, and continue from the primary through all subsequent iterations. Moreover, ANY sequence of numbers that begins and ends with the same number has increments that sum to zero. Adding zeros before and after any binaric increases in value as it moves deeper into intra-space.

Increment Analysis and the subsequent unveiling of hidden number patterns is interesting in itself. Perhaps in the future this ***intra-space math*** may find a niche in an area that has not yet been discovered. In the meantime, the beauty of numbers and the negation of the notion of randomness is exciting. Increment analysis opens up a whole new set of options for analyzing numbers.  New unsuspected relationships and hidden patterns become apparent that are beautifully symmetrical.

# Appendix

# INCREMENT ANALYSIS SAMPLE WORKSHEET

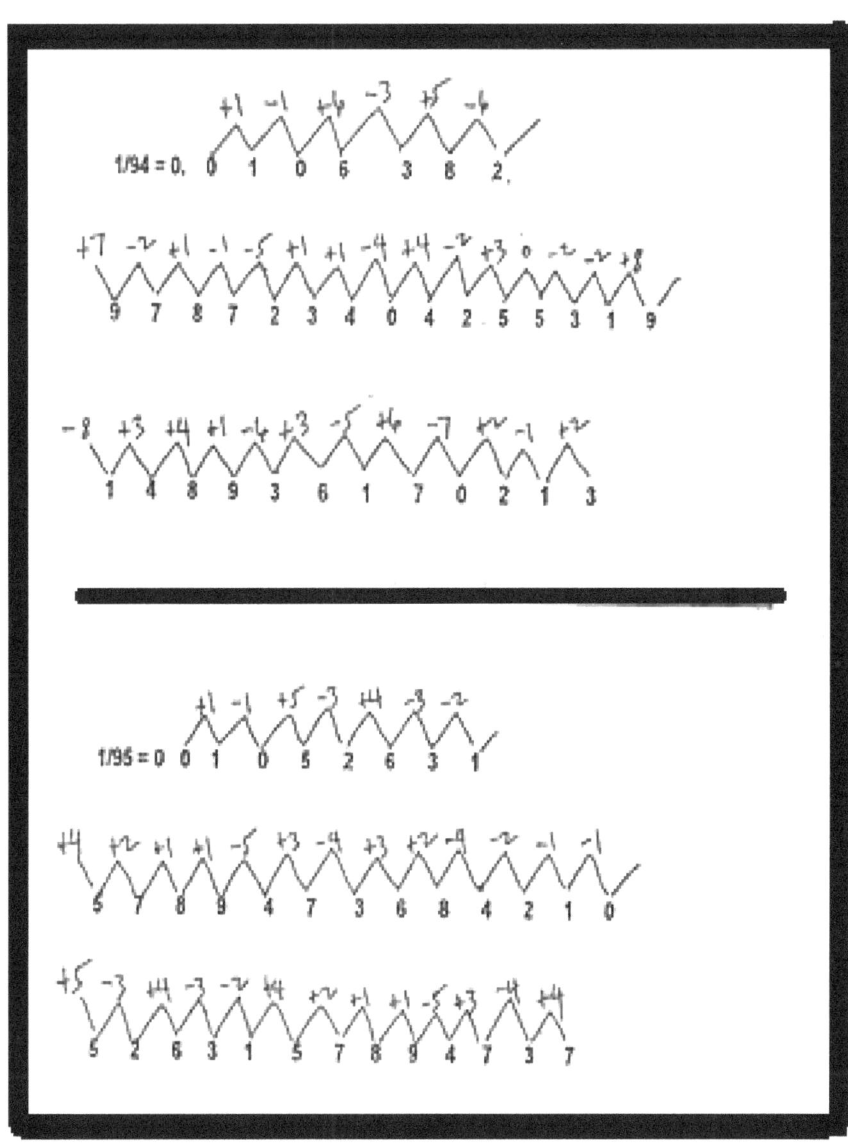

# THE "GENDER" OF NUMBER SEQUENCES

Though it is true that all number sequences have binaries that sum to zero if zeroes are added before and after the sequence, it is also true that a certain type of number sequence automatically sums to zero, i.e. FEMALE number sequences that are represented by circular numbers.

Example of a *Feminine* Sequence:

| | | |
|---|---|---|
| 1/23 = 0.04347826086956521739 130... | +4 -1 +1 +3 +1 -6 +4 −6 +8 -2 +3<br><br>-4 +1 -1 -3 -1 +6 -4 +6<br><br>-8 +2 -3 | *22-element mirror*<br><br>*+39 -39* |

Example of a *Masculine* Sequence:

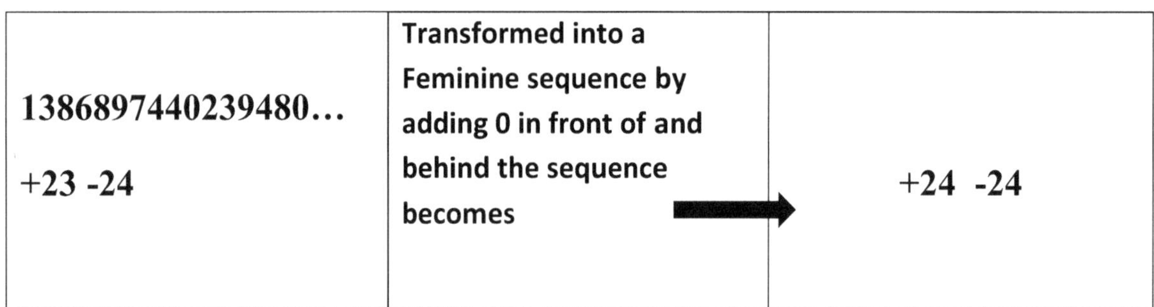

| | | |
|---|---|---|
| **1386897440239480…**<br><br>**+23 -24** | **Transformed into a Feminine sequence by adding 0 in front of and behind the sequence becomes** | **+24  -24** |

***Fundamentally, all "Masculine"sequences are comprised of a series of "Female" sequences.*** For example, when you have a lengthy number string, you will come upon a number that previously occurred in the sequence. When that occurs, the numbers between the initial number and the next number of the same kind sum to zero. This represents a "circle."

# MIRROR SEQUENCES

## (For reciprocals of circular numbers between 1 – 100)

| | | |
|---|---|---|
| 1/7 = 0.142857142857… | +3  -2  +6   -3  +2  -6 | **Six-element mirror** |
| 1/13 = 0.0769230769230… | +7  -1  +3   -7  +1  -3 | **Six-element mirror** |
| 1/14 = 0.07142857142857… | (+7) –6  +3  -2  +6  -3  +2 … | **Six-element mirror** |
| 1/17 = 0.05882352941176470 | +5  +3  0  -6  +1  +2  -3  +7<br><br>-5  -3  0  +6  -1  -2  +3  -7 | *16-element mirror* |
| 1/19 = 0.0526315789473684210… | +5  -3  +4  -3   -2  +4  +2  +1  +1<br><br>-5  +3  -4  +3  +2   -4  -2   -1  -1 | *18-element mirror* |

| | | |
|---|---|---|
| 1/23 =<br>0.04347826086956521739<br>130… | +4 -1 +1 +3 +1 -6 +4 –6 +8<br><br> -2 +3<br><br>-4 +1 -1 -3 -1 +6 -4 +6 -8 +2 -3 | *22-element mirror* |
| 1/26 =<br>0.03846153846153846153<br>846… | ( +3 ) +5 -4 +2   -5 +4 -2 +3 –4 +2 | *6-element mirror* |
| 1/28 =<br>0.0357142857142857… | (+3 +2) +2 -6 +3  -2 +6 -3 | *6-element mirror* |
| 1/29 =<br>0.03448275862068965517<br>241379310… | +3 +1 0 +4 -6 +5 -2 +3 -2<br><br> -4 -2 +6 +2 +1<br><br>-3 -1 0 -4   +6 -5 +2 -3 +2 +4 +2 -6 -2 -1 | *28-element mirror* |
| 1/34 =<br>0.02941176470588235… | (+2 +7) -5 -3  0<br><br>+6 -1 -2 +3 -7 +5 +3  0<br><br>-6 +1 +2 -3 +7 -5 -3  0 | *16-element mirror;*<br><br>*Oscillating –5 -3 0, +5 +3 0* |

| 1/35 =<br>0.02857142857142857… | (+2)  +6  -3  + 2  -6  +3<br><br> -2  +6 -3… | **6-element mirror** |
| 1/38 =<br>0.02631578947368421052<br>63… | (+2)<br><br> +4  -3  -2  +4  +2  +1  +1  -5  +3<br><br> -4  +3  +2  -4  -2  -1  -1  +5  -3  (+4  -3) | **18-element mirror** |
| 1/39 = 0.025641025641… | +2  +3  +1  -2  -3  -1… | **6-element mirror**<br><br>**Ps = 0** |
| 1/46 =<br>0.02173913043478260869<br>5652… | -1  +6  -4  +6  -8  +2  -3  +4  -1  +1  +3<br><br><br>+1  -6  +4  -6  +8  -2  +3  -4  +1 -1  -3 | **22-element mirror**<br><br>**Ps = 0** |
| 1/47 =<br>0.02127659574468085106<br>38297872340425… | +2  -1  +1  +5  -1  -1  +4  -4  +2<br><br> −3  0  +2  +2  -8  +8  -3  -4  -1  +6  -3  +5  -6  +7<br><br><br>-2  +1  -1  -5  +1  +1  -4  +4  -2  +3  0  -2  -2  +8  -8  +3  +4  +1<br><br> -6  +3 −5  +6  -7 | **46-element mirror**<br><br>**Ps = 0** |

| | | |
|---|---|---|
| 1/49 =<br>0.02040816326530612244<br>89795918367346… | +2  -2  +4  -4  +8  -7+5  -3  -1 +4 −1 -2  -3  +6  -5  +1  0 +2  0  +4  +1<br><br>-2  +2  -4  +4  -8  +7  -5  +3 +1 −4<br><br>+1  +2  +3  -6  +5  -1  0  -2 0  -4  -1 | *42-element mirror*<br><br>*Ps = 0* |
| 1/52 =<br>0.019230769230769230… | ( +1  +8)  -7  +1  -3<br><br>      +7  -1  +3… | **6-element mirror** |
| 1/56 = 0.017857<br><br>142857 142857… | (+1  +6  +1)  -3  +2  -6<br><br>      +3  -2  +6… | *Six-element mirror* |
| 1/58 =<br>0.01724137931034482758<br>620689655  17241… | (+1)<br>+6  -5  +2  -3  +2  +4  +2  -6  -2<br><br>-1  +3  +1  0  +4<br><br><br>-6  +5  -2  +3  -2  -4  -2 +6  +2  +1  -3  -1  0  -4… | *28-element mirror* |

| | | |
|---|---|---|
| **1/59 =**<br>**0.01694915254237288135**<br>**59322033898305…** | **(+1)**<br><br>**+5  +3  -5  +5  -8  +4  -3  +3  -1**<br><br>**-2  +1  +4  -5  +6  0  -7  +2  +2  0  +4  -6  -1  0  -2  +3  0  +5  +1  -1**<br><br><br>**-5  -3  +5  -5  +8  -4  +3  -3  +1**<br><br>**+2  -1  -4  +5  -6  0  +7  -2  -2  0**<br><br>**-4  +6  +1  0  +2  -3  0  -5  -1  +1** | *58-element mirror* |
| **1/61 =**<br>**0.01639344262295081967**<br>**21311475409836…** | **+1  +5  -3  +6  -6  +1  0  -2  +4  -4  0  +7  -4  -5  +8  -7  +8  -3  +1  -5  -1  +2  -2  0  +3  +3  -2  -1  -4  +9**<br><br>**-1  -5  +3……** | *60-element mirror?* |

| | | |
|---|---|---|
| 1/63 = 0.015873  015873 015873  015873 015873 0158… | +1  +4  +3   -1 - 4  -3 | *Six-element mirror* |
| 1/65 = 0.0153846  153846 153846153846 153846 153… | (+1)<br><br>+4  -2  +5   -4  +2  -5 | *Six-element mirror* |
| 1/67 = 0.01492537313432835820 8955223880597… | (+1) +3  +5  -7  +3  -2  +4  - 4  -2  +2  +1  -1  -1  +6  -5 +2  +3  -6  -2  +8  +1  -4  0 -3  0  +1  +5  0<br><br> -8  +5  +4  -2 … | *66-element mirror…* |
| 1/68 = 0.014705882352941176 4705882352941176… | (+1  +3)<br><br>+3  -7  +5  +3  0  -6  +1 (+2)<br><br>-3  +7  -5   -3  0  +6   -1 | *14-element mirror* |

| | | |
|---|---|---|
| 1/70 = 0.0142857 142857142857… | (+1)  +3  -2  +6   -3  +2  -6 | *6-element mirror* |
| 1/73 = 0.013698630 13698630 13698630 136986301… | +1  +2  +3  +3<br><br>-1    -2  -3  -3 | *8-element mirror* |
| 1/76 = 0.01315789473684210526 31578947368421… | (+1)   +2<br><br>-2  +4  +2  +1  +1  -5  +3  -4  -3<br><br>+2  -4  -2    -1  -1  +5  -3  +4  -3 | *18-element mirror* |
| 1/77 = 0.0129870  129870 129870  129870  129870… | +1  +1  +7    -1  -1  -7 | *6-element mirror* |
| 1/85 = 0.01176470588235294 1176470588235294… | (+1)<br>0  +6  -1   -2   +3  -7  +5  +3<br>0   -6  +1  +2  -3  +7  -5    -3 | *16-element mirror* |

| 1/89 = 0.01123595505617977528 08988764044944... | +1  0 +1  +1  +2  +4  -4  0 -5  +5  +1  -5  +6  +2  -2  0 -2 –3  +6  -8  +8  +1<br><br>-1  0  -1  -1  -2  -4  +4  0  +5<br><br> -5  -1  +5  -6  -2  +2  0  +2  +3  -6  +8  -8  -1** | *44-element mirror* |
| --- | --- | --- |
| **1/91 = 0.010989 010989 010989 010989 010989...** | +1  -1  +9  -1  +1  -9 | *6-element mirror* |
| 1/92 = 0.01086956521739130434 78260869565217 | (+1  -1)  +8  -2  +3  -4  +1 -1  -3  -1  +6  -4  +6<br><br>-8  +2  -3  +4  -1  +1  +3  +1  -6  +4  -6 | *22-element mirror* |
| 1/95 = 0.01052631578947368421 05263157894737 | (+1  -1)  +5  -3  +4  -3  -2  +4  +2  +1  +1<br><br><br>-5  +3  -4  +3  +2  -4  -2  -1  -1 | *18-element mirror* |
| 1/98 = 0.01020408163265306122 44897959183673... | (+1  -1)  +2  -2  +4  -4  +8  -7  +5  -3  -1  +4  -1  -2  -3  +6  -5  +1  0  +2  0  +4  +1<br><br>-2  +2  -4  +4  -8  +7  -5  +3  +1  -4... | *42-element mirror* |

# BIBLIOGRAPHY

Dunham, William. *The Mathematical Universe*. John Wiley & Sons, Inc. New York. 1994.


King, Jerry P. *The Art of Mathematics*. Fawcett Columbine. New York. 1992.

Schwaller de Lubicz. R.A. *A Study of Numbers*. Inner Traditions International. Rochester, Vermont. 1986.


Bowen, Joyce P. *Unimath: Hidden Supersymmetry in All Number Sequences.* Astropoint Publishing. Chicago. 2012.


Fuller, R. Buckminster. Applewhite, E.J. *Synergetics: Explorations in the Geometry of Thinking.* Macmillan Publishing Co. New York. 1975.


Balmond, Cecil. *Number 9: The Search for the Sigma Code*. Prestel. Munich-New York. 1998.

Blatner, David. *The Joy of Pi*. Walker and Company. New York. 1997.

Georges, Ifrah. *The Universal History of Numbers*. John Wiley & Sons, Inc. New York. 2000.


Merrill, Helen A. *Mathematical Excursions*. Dover Publications, Inc. U.S.A. 1933, 1957.

Clapham, Christopher. *The Concise Oxford Dictionary of Mathematics,* 2nd Edition. Oxford University Press. Oxford-New York. 1996.

Boyer, Carl B. Merzbach, Uta C. *A History of Mathematics* 2nd Edition. John Wiley & Sons, Inc. New York. 1991.

Wells, David. *The Penguin Dictionary of Curious and Interesting Numbers.* Penguin Books. London, England. 1986.

Ogilvy, Stanley C. Anderson, John T. *Excursions in Number Theory.* Dover Publications, Inc. New York. 1988.

Pickover, Dr. Clifford A. *The Zen of Magic Squares, Circles, and Stars.* Princeton University Press. 2001.

Plichta, Peter. *God's Secret Formula: Deciphering the Riddle of the Universe and the Prime Number Code.* Element Books. Rockport, MA. 1997.

Bennett, Deborah J. *Randomness.* Harvard University Press. Cambridge, MA. 1998.

Gilles, William F. *The Magic and Oddities of Numbers.* Vantage Press, Inc. New York. 1953.

Camm, F. J. *Mathematical Tables and Formulae.* Philosophical Library. New York. 1958.

Watkins, Matthew. *Useful Mathematical & Physical Formulae.* Walker & Company. New York. 2000.

Dwight, Herbert Bristol. *Mathematical Tables of elementary and some higher mathematical functions.* Dover Publications, Inc. New York. 1961.

Saradar, Ziauddin. Ravetz, Jerry. Van Loon, Borin. *Introducing Mathematics.* Totem Books. New York. 1999.

Salem. Testard. Salem. *The Most Beautiful Mathematical Formulas.* John Wiley and Sons, Inc. 1992.

Dunham, William. *Journey Through Genius (The Great Theorems of Mathematics).* Penguin Books, John Wiley & Sons. 1990.

Oyibo, Gabriel. *Grand Unified Theorem*. Nova Science Publishers. New York. 2001.

Bowen, Joyce P. *The Law of Digit Balance*. Astropoint Publishing. Chicago, 2003.

Holroyd, Stuart, Ed. *The Arkana Dictionary of New Perspectives*. London. The Penguin Group. 1989.

***INTERNET RESOURCES***

Gamboa, Ana. Alpha – *The Fine Structure Constant*
http://www.physicspost.com/articles.php?articleId=11

*Can Negative Numbers be Prime?* Prime Pages FAQ

http://www.utm.edu/research/primes/notes/faq/negative_primes.html

Michael Conrad Tilstra (Tadpol) Basic Infinite Math. http://tadpol.org/theory/infinitemath.html

Mathworld. Transcendal Number.

http://mathworld.wolfram.com/TranscendentalNumber.html

Pascal's Triangle. http://www.csam.montclair.edu/~kazimir/patterns.html

Ancient Light and Uncertain Theories.

http://www.geocities.com/syzygywjp/AncientLight.html

Sirag, Saul-Paul. *The World as Cryptogram*. International Space Sciences Organization.

November 7, 2000.

The First 100 Fibonacci Numbers.

http://math.holycross.edu/~davis/fibonacci/fib0-99.html

Riemann Zeta Function – MathWorld

http://mathworld.wolfram.com/RiemannZetaFunction.html

The Riemann Hypothesis.  Prime Pages.

http://www.utm.edu/research/primes/notes/rh.html

*When constants are not constant*. Physics Web. October, 2001.

http://physicsweb.org/article/world/14/10/4

Rice, Aaron. *Infinity, The Truth About*. 1997/02/10

http://www.galactic-guide.com/articles/8R69.html

*The First 1000 Primes.*

http://www.utm.edu/research/primes/lists/small/1000.txt

The First 200 Lucas Numbers and their factors.

http://www.mcs.surrey.ac.uk/Personal/R.Knott/Fibonacci/lucas200.html


Devlin, Keith. *How Euler discovered the zeta function.*

http://www.maa.org/devlin/Zeta.PDF


*The 23 Paris Problems.* http://www.math.umn.edu/~wuttnab/problems2.html


*Latin Squares*. http://lwww.cut-the-knot.org/arithmetic/latin.shtml


Ivars Peterson's MathTrek. *The Limits of Mathematics.*


http://www.sciencenews.org/sn_arc98/2_21_98/mathland.htm

*Magic Square* – from MathWorld.

http://mathworld.wolfram.com/MagicSquare.html


The Math Forum.  *Why are Operations of Zero so Strange?*

http://mathforum.org/library/drmath/view/55764.html


*Re: Infinity*.  http://www.math.tulane.edu/bulletin/messages/100.html

Michael Conrad Tilstra (Tadpol) Zero over Zero.  http://tadpol.org/theory/zero.html

*Complex Infinity*.  http://www.mupad.com/doc/eng/stdlib/cinfty.shtml

# NOTES

# NOTES

# NOTES

# NOTES

# NOTES

# NOTES

# NOTES

# ABOUT THE AUTHOR

Dr. Joyce P. Bowen, PhD, is a life-long educator, polymath, and author or editor of over 20 books and articles. Among her notable works are *The Law of Digit Balance; Uniphysics: the Science of Synthesis; Unimath: Hidden Super Symmetry in All Number Sequences; Ben Franklin's Most Magical Magic Square: New Analysis; and Starring the Circle: the Circle-lation of Numbers.* Dr. Bowen holds a BS in Education from Chicago State University; an MA in Fine and Performing Arts from Governors State University, and a PhD in Holistic Medicine from the Union Institute and University. She also holds certificates in Manuscript Editing, Book Publishing, and Publication Design from the University of Chicago. Dr. Bowen resides in Chicago, Illinois.